Kenneth Alvin Solomon
Wei-Kuang Chao
Jesse Kendall

Defining the Criteria for Air Bag Activation in Passenger Vehicles

Kenneth Alvin Solomon
Wei-Kuang Chao
Jesse Kendall

Defining the Criteria for Air Bag Activation in Passenger Vehicles

LAP LAMBERT Academic Publishing

Impressum / Imprint
Bibliografische Information der Deutschen Nationalbibliothek: Die Deutsche Nationalbibliothek verzeichnet diese Publikation in der Deutschen Nationalbibliografie; detaillierte bibliografische Daten sind im Internet über http://dnb.d-nb.de abrufbar.

Alle in diesem Buch genannten Marken und Produktnamen unterliegen warenzeichen-, marken- oder patentrechtlichem Schutz bzw. sind Warenzeichen oder eingetragene Warenzeichen der jeweiligen Inhaber. Die Wiedergabe von Marken, Produktnamen, Gebrauchsnamen, Handelsnamen, Warenbezeichnungen u.s.w. in diesem Werk berechtigt auch ohne besondere Kennzeichnung nicht zu der Annahme, dass solche Namen im Sinne der Warenzeichen- und Markenschutzgesetzgebung als frei zu betrachten wären und daher von jedermann benutzt werden dürften.

Bibliographic information published by the Deutsche Nationalbibliothek: The Deutsche Nationalbibliothek lists this publication in the Deutsche Nationalbibliografie; detailed bibliographic data are available in the Internet at http://dnb.d-nb.de.

Any brand names and product names mentioned in this book are subject to trademark, brand or patent protection and are trademarks or registered trademarks of their respective holders. The use of brand names, product names, common names, trade names, product descriptions etc. even without a particular marking in this work is in no way to be construed to mean that such names may be regarded as unrestricted in respect of trademark and brand protection legislation and could thus be used by anyone.

Coverbild / Cover image: www.ingimage.com

Verlag / Publisher:
LAP LAMBERT Academic Publishing
ist ein Imprint der / is a trademark of
OmniScriptum GmbH & Co. KG
Heinrich-Böcking-Str. 6-8, 66121 Saarbrücken, Deutschland / Germany
Email: info@lap-publishing.com

Herstellung: siehe letzte Seite /
Printed at: see last page
ISBN: 978-3-659-69798-2

Copyright © 2015 OmniScriptum GmbH & Co. KG
Alle Rechte vorbehalten. / All rights reserved. Saarbrücken 2015

TABLE OF CONTENTS

Chapter 1	**Introduction**	1
	What Are Air Bags Designed to Do?	2
	Effectiveness of Air Bags	4
	References	6
Chapter 2	**Steps Involved in an Air Bag Deployment**	7
	Crash Sensors	9
	The "Black Box" – Event Data Recorder	10
	The Decision Process	12
	Algorithm Variations	15
	"Smart" Air Bags	17
	References	18
Chapter 3	**Defining the Thresholds**	20
	Threshold Estimates	21
	References	24
Chapter 4	**Comparing Values**	25
	Case Study #1 – 2007 Cadillac Escalade	28
	Case Study #2 – 2007 Ford Edge	30
	Case Study #3 – 1999 GMC Jimmy	32
	Case Study #4 – 2008 Scion tC	34
	Case Study #5 – 2012 Ford F550	36
	Case Study #6 – 2006 Cadillac CTS	38
	Case Study #7 – 2007 Chevrolet Equinox	40

	Case Study #8 – 2007 Chevrolet Corvette	42
	Summary	44
	References	45
Chapter 5	**Conclusion**	**46**
Appendix	**Glossary of Terms**	**49**

CHAPTER 1

INTRODUCTION

Ask anyone, including the manufacturer of your vehicle, "When will my air bag deploy?" and you may not get a straight answer. With the race to protect passengers comes the ambiguity, complexity, and protection of privacy of the vehicle manufacturer's intellectual property. To further complicate matters, air bag deployment systems have changed rapidly and will continue to change with new technology. In forensic investigations of automobile accidents, a large percentage of injuries caused by vehicle air bag deployment can be traced to deficiencies in the sensing system (Cheong et al., 2012, p.1). Vehicle damage estimates used to determine vehicle speed or body acceleration at the time of injury are often unreliable. Crash data from the vehicle's "black box" retrieval system may be either invalid or even compromised at times. For these reasons, it is important to understand the air bag system and deployment criteria and how it can be used as another forensic analysis tool.

This book will provide an introduction to air bag components, followed by a step-by-step breakdown of the deployment process. Variables used in air bag deployment algorithms are discussed, and examples of several patented systems are compared. A method to estimate the range of speed, deceleration, or displacement thresholds for air bag deployment will also be provided. In addition, case studies are presented to compare the predicted results (based on the estimates) and the actual outcome. The objective of this book is to provide the readers an understanding of the air bag system and components, as well as a scientific approach to determine when an air bag should or should not deploy.

What are Air Bags designed to do?

According to Federal Motor Vehicle Safety Standards 208 Section 4.1.5.3, all passenger vehicles manufactured on or after September 1, 1997, must be equipped with an air bag for the driver and right front passenger (US Department of Transportation, 1998). Air bags are designed to be a Supplemental Restraint System (SRS), mostly to provide supplemental protection for belted front-seat occupants in frontal crashes; air bags are not intended as the only form of occupant protection.

In the event of a frontal crash collision, seat belts act as primary occupant restraints; however, the force due to deceleration is applied only over the width of the seat belt, resulting in unequal pressure distribution across the occupant's body and the potential for injury. As a supplemental restraint system, air bags exert nearly equal pressure across the entire contact area with the occupant's body in accordance with the formula of pressure (i.e., *Pressure = Force/Area*). Given a similar impact force for occupants restrained by seat belts versus occupants restrained by air bags, it is clear that the air bag provides a larger contact area over which the force can be distributed; therefore, the maximum pressure experienced by the occupant's body is vastly reduced with the use of air bags.

The purpose of the air bag is to provide a cushion between the occupants and the vehicle's interior. In order for air bags to be effective, they have to be fully inflated to their designed shape in a short amount of time, before the occupants make contact with them; the combination of this rapid inflation and the speed of impact can potentially cause serious or fatal injuries to certain people if they are in contact with the air bag during its inflation process (Struble 1998). Therefore, air bags must have a control system that can recognize a crash correctly, and this must take place early enough for the air bags to inflate safely.

Figure 1.1 (Tate, 2010) below illustrates an ideal scenario in which the air bag deploys and becomes a cushion to mitigate the secondary impact between the belted driver and the steering wheel area.

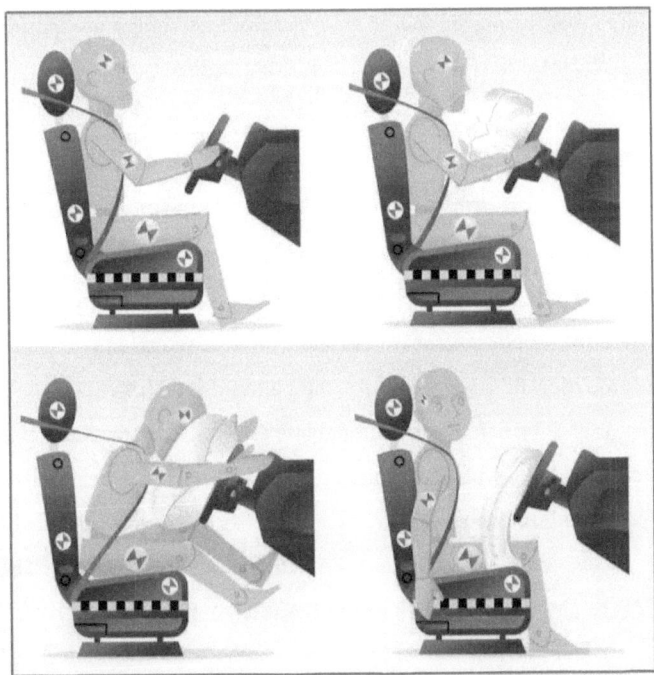

Figure 1.1   How a driver's air bag deployment protects the belted driver

For a crash in which the entire front of the vehicle is impacted, the bag typically inflates within about 1/20 of a second after impact. The inflated air bag creates a protective cushion between the occupant and part of the vehicle's interior (i.e., steering wheel, dashboard, and windshield). At about 4/20 of a second following impact, the air bag begins to deflate. The entire deployment, inflation, and deflation cycle is over in less than one second. After deployment, the air bag deflates rapidly as the gas escapes through vent holes or through the air bag's porous nylon fabric. Initial deflation enhances the cushioning effect of the air bag by maintaining approximately the same internal pressure as the occupant strokes into the bag. Subsequent rapid and total deflation enables the driver to maintain control if the vehicle is still moving after the crash and ensures that the driver and/or the right-front passenger are not trapped by the inflated air bag (NHTSA 2001).

Effectiveness of Air Bags

According to the statistics published by National Highway Traffic Safety Administration (NHTSA), air bags and lap-shoulder belts when used together have an estimated 51 percent fatality-reducing effectiveness. Air bags provide (coupled with seat belts) about an 11 percent reduction in fatality risk for the belted driver (relative to a belted driver without air bags), and 14 percent for the unbelted driver in all crashes as compared to identical vehicle accidents where air bags are not used, but only seat belts are utilized. Concerning overall injury reduction for drivers, for serious injury, the air bag plus lap-shoulder belt use provide 68 percent reduction in injury risk. The estimated effectiveness of the air bag alone is 30 percent (NHTSA 2001).

NHTSA estimates that air bags saved 8,369 lives between 1987 and 2001 (Glassbrenner, p.1). Between 2008 and 2012, air bags saved 11,682 lives (NHTSA 2013, p.1). Figure 1.2 below is a table of numbers extracted from data published by NHTSA, pertaining to the effectiveness of seat belts and air bags. Note again that air bags are considered as a Supplemental Restraint System (SRS), rather than as a replacement of the existing seat belts. For the data presented in Figure 1.2 below, the use of air bags is accompanied by the use of seat belts.

Year	Lives Saved, Age 4 & Younger	Lives Saved, Age 5 & Older	Lives Saved, Age 13 & Older
	Child Restraints	Seat Belts Alone No Air Bags	Frontal Air Bags With Seat Belt Use
2008	286	13,312	2,557
2009	307	12,763	2,387
2010	303	12,582	2,315
2011	262	11,983	2,210
2012	284	12,174	2,213

Figure 1.2   NHTSA data of lives saved by restraint use, 2008 to 2012

Seat belts may be the most effective safety restraining devices to protect the occupant(s) of a vehicle in a collision; though, as seen in Figure 1.2, air bag use has proven to further enhance occupant protection and saved additional lives by approximately 18 to 19 percent, which is a significant increase.

References

- Cheong, Denove, Rowell & Bennett. (2012). *Electronic Crash Sensing in Air Bag Litigation.* Society of Automotive Engineers, Inc.
- S. Ferguson, A. Lund & M. Greene. (1995). *Driver Fatalities in 1985-1994 Air Bag Cars.* NHTSA.
- D. Glassbrenner. *Estimating the Lives Saved by Safety Belts and Air Bags.* NHTSA Paper No. 500.
- R. Tate. (2010). Figure retrieved from http://www.istockphoto.com/vector/air-bag-crash-dummies-14029108.
- National Highway Traffic Safety Administration. (November 2001). *Effectiveness of Occupant Protection Systems and Their Use – DOT HS 809 442.*
- National Highway Traffic Safety Administration. (2013, November). *Lives Saved in 2012 by Restraint Use and Minimum Drinking Age Laws.* Traffic Safety Facts. Retrieved from http://www-nrd.nhtsa.dot.gov/Pubs/811851.pdf.
- D. Struble. (1998). *Airbag Technology: What it is and How it Came to Be.* SAE International.
- U.S. Department of Transportation. (1998). *Occupant Crash Protection – Standard No. 208.* Federal Motor Vehicle Safety Standards and Regulations.

CHAPTER 2

STEPS INVOLVED IN AN AIR BAG DEPLOYMENT

In general, an air bag assembly includes the following components: the crash sensor(s), the air bag control module (ACM), and the actual air bag module. The crash sensor(s), located on the front of the vehicle or in the passenger compartment measure deceleration, the rate at which a vehicle slows down. The ACM is an electronic device that monitors the operational readiness of the air bag system whenever the vehicle ignition is turned on and while the ignition is powered. The air bag module, containing an inflator and a vented or porous lightweight fabric air bag, is located in the hub of the steering wheel on the driver side or in the instrument panel on the passenger side (NHTSA 2001). Figure 2.1 below depicts the components of air bag systems.

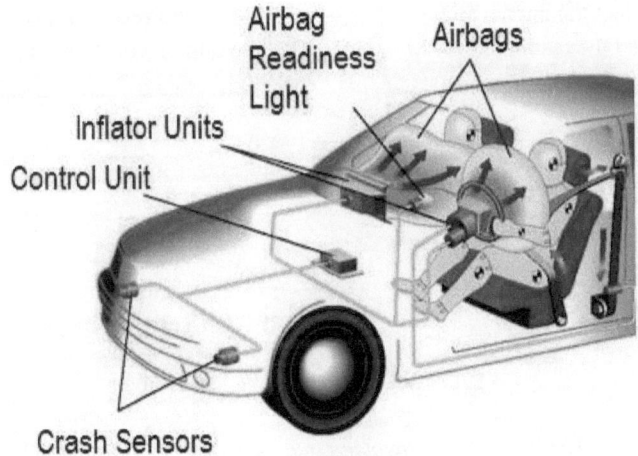

Figure 2.1 Components of Air Bag Systems

In the event of a frontal collision, the crash sensor(s) will detect the impact and provide its reading to the air bag control module; after a decision process based on the implemented algorithm, a "fire" or "no fire" command will be issued. If the detected impact qualifies for a deployment (i.e., if a "fire" command is issued), an electric current from the air bag control module is sent to the inflator unit(s) within the air bag module. This detonator starts a chemical reaction, producing nitrogen gas that rapidly inflates the nylon fabric air bag. Figure 2.2 below is a brief flowchart illustrating the steps involved in an air bag deployment.

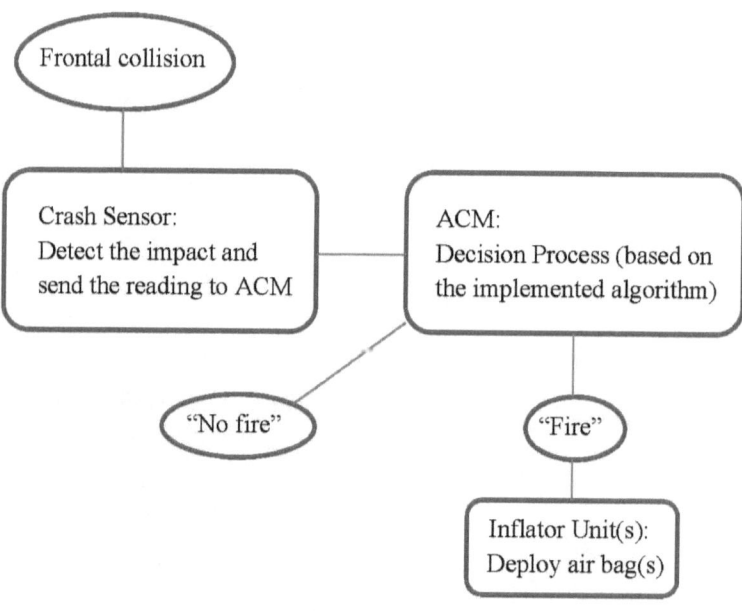

Figure 2.2 Steps Involved in an Air Bag Deployment

Crash Sensors

Early air bag deployment systems in older vehicles utilized mechanical sensors for crash detection, which were phased out of the US market around 1994 (Rowell 2012). Early mechanical sensors, such as the "Rolamite" by Sandia National Laboratories, relied on a metallic sphere that was stabilized at a standby position by a spring or a magnet (see Figure 2.3 below).

When the sensor was subjected to a force beyond a designed threshold, the spring or magnet could no longer keep the metallic mass in place. The mass moved and made contact with an electrode, sending an electrical signal to the air bag control module. Systems with mechanical sensors were generally inaccurate at interpreting minor collisions, causing unnecessary deployments. Movement within mechanical sensors can be underrepresented with frontal collisions, and the acceleration experienced by the sensor is sometimes slightly delayed, so air bags do not deploy or deploy with an undesired delay. As the result of improvements, modern air bag deployment now relies on microelectromechanical system (MEMS) components.

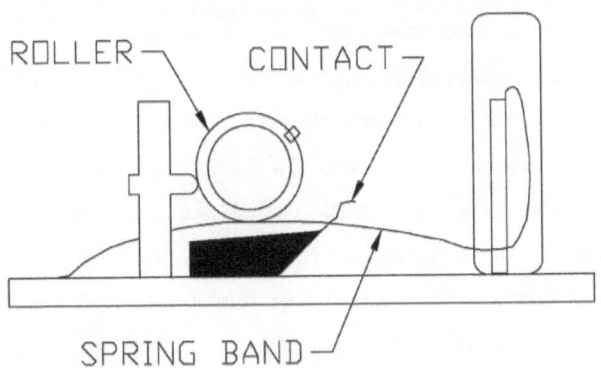

Figure 2.3   Mechanical Air Bag Sensor

MEMS crash sensors measure acceleration with an accelerometer that sends a continuous stream of data to the air bag control module. Accelerometers are typically piezoelectric or variable capacitance sensors. The most common MEMS accelerometer in use today is the ADXL-50 by Analog Devices (see Figure 2.4 below). As an anchored mass moves relative to the sensor's body due to acceleration, a plate attached to the anchored mass moves closer to a stationary plate. The change in distance between the plates affects the capacitance of the sensor, or the ability to hold an electrical charge. This change in capacitance is measured and then converted to a change in voltage. The voltage change is directly correlated to force due to acceleration, and the readings are interpreted as acceleration by the air bag control module. Using an algorithm, the control module can determine if air bag deployment is necessary based on the pattern of the acceleration pulses over time. With the development of the MEMS sensors, the issues with undesired air bag deployment and non-deployment have been reduced.

The "Black Box" – Event Data Recorder

The involvement of NHTSA in the development of automotive crash recorders began in 1969 with an intent to measure the acceleration and time of the crash only. It was later recognized that crash avoidance research and automakers may benefit from the pre-crash data and the acceleration/time history (Klaber 1981). As the Event Data Recorder (EDR) was developed to detect a crash and record the vehicle's motion and the driver's maneuvering (and even the use or non-use of the seat belt) during a pre-defined time period before and after the accident, it has been concluded that analytical considerations should be taken when designing pre-filtering circuits and selecting appropriate parameters for identifying crash accidents (Lee 2004, Gabler et al. 2004). Based on the crash tests from 2006 to 2009 in Japan, the pre-crash velocity and the maximum delta-V recorded in the EDRs, when compared with the results from accelerometers and high speed video cameras, showed good reliability and accuracy (Takubo, et al. 2011).

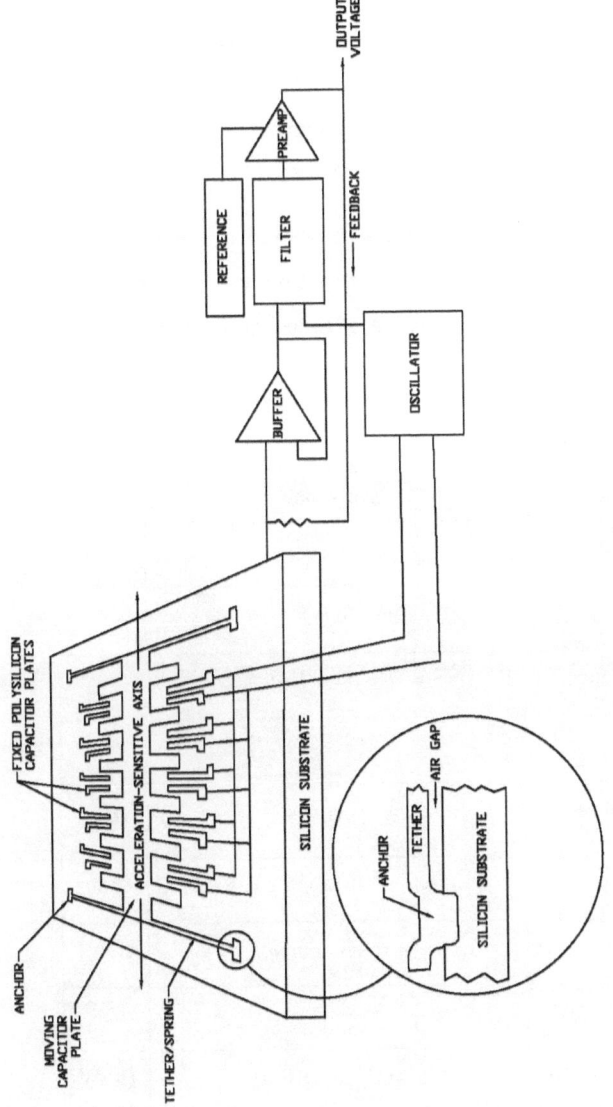

Figure 2.4   Microelectromechanical Sensor (MEMS)

The Decision Process

The air bag control module (ACM) receives a continuous signal from each MEMS sensor and records the data for a fixed period after a specific event. With a central processing unit (CPU), it performs algorithmic computations and controls the "fire" or "no-fire" command for air bag deployment.

The triggering algorithms determine crash severity by evaluating one or more of the kinematic variables, which may include, but are not limited to, displacement, velocity, acceleration, "jerk" (defined as the derivative of acceleration over time), etc. A list of the kinematic variables is listed in Table 2.1 below.

Variable	Expression	Units
Acceleration	$a = \dfrac{dv}{dt}$	ft/s²
Velocity	$v = \int a\,dt = \dfrac{dx}{dt}$	ft/s
Displacement	$x = \int v\,dt = \iint a\,dt\,dt$	ft
Jerk	$j = \dfrac{da}{dt}$	ft/s³
Energy Density	$e = \int_{x_0}^{x} a\,dx = \int_{v_0}^{v} v\,dv = \dfrac{1}{2}(v^2 - v_0^2)$	(ft/s)²
Energy	$E = \dfrac{1}{2} m(v^2 - v_0^2), \quad m: mass$	ft-lb
Power	$\bar{p} = \dfrac{dE}{dt} = mva$	ft-lb/s
Power Density	$p = \dfrac{\bar{p}}{m} = va$	ft²/s³
Power Rate Density	$p' = \dfrac{dp}{dt} = vj + a^2$	(ft/s²)²

Table 2.1 Kinematic variables used in air bag triggering algorithms.

In general, among the kinematic variables, three are examined: the change in velocity, vehicle speed, and vehicle acceleration. Change in velocity is used to detect crash, vehicle speed is used to determine airbag deployment, and acceleration information determines the crash severity (Hussain, et al. 2006). Examples of algorithm decision flow charts are shown in the following Figures 2.5, 2.6, and 2.7.

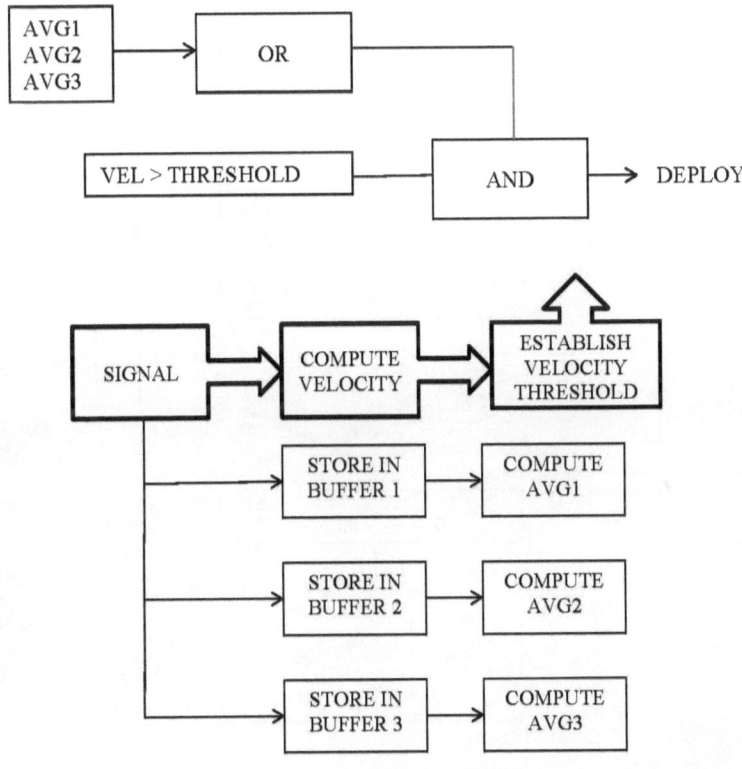

* AVG1, AVG2, AVG3 = average acceleration over 4, 8 and 24 milliseconds, respectively

Figure 2.5 Algorithm Flow Chart Using '$\bar{a}$' and 'Δv' Variables (McConnell 2001)

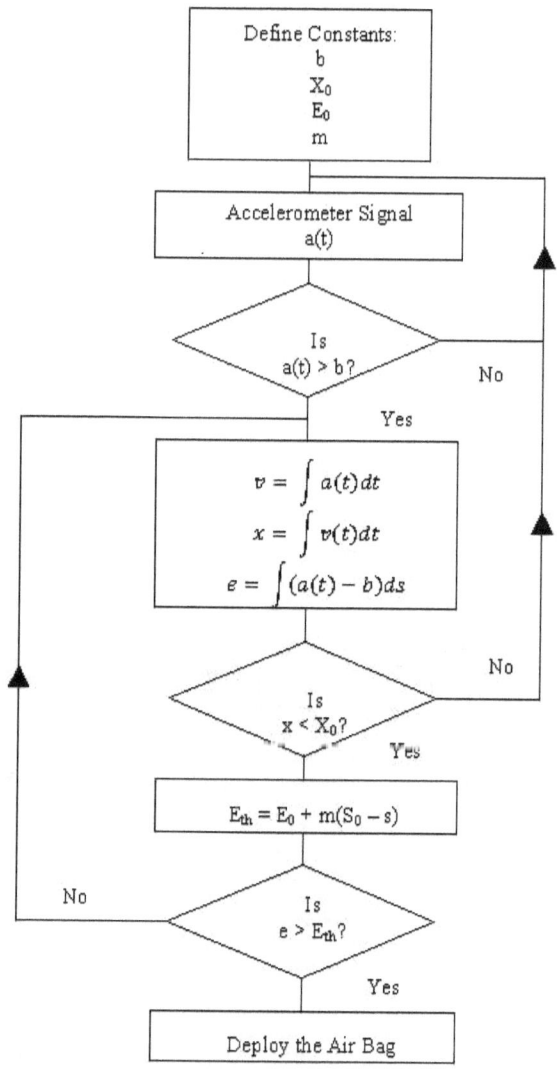

Figure 2.6 Algorithm Flow Chart using 'Δv', 'x', and 'e' Variables (Huang 2002)

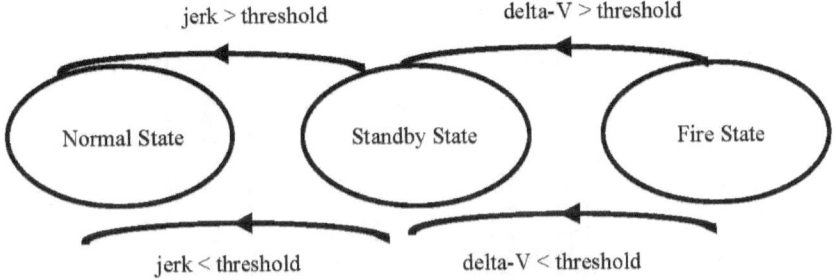

Figure 2.7 Algorithm Flow Chart Using 'jerk' and 'Δv' Variables (Hussain, et al. 2006)

ALGORITHM VARIATIONS

Crash sensing schemes vary greatly between patents. A majority of systems patented after 1995 utilize *delta-v, acceleration,* or *jerk,* as variables in the system wakeup command, and for triggering the air bag. Note that air bag deployments are not solely dependent on these kinematic variables. Recent systems also include *occupant sensing* and analysis of the *distance* from the occupant (Breed, et al. 2000). A manufacturer may choose, for example, to not deploy the passenger air bag if there is no occupant seated in the right front passenger location (Gabler, et al. 2008).

Table 2.2 below outlines the algorithms used between 1995 and 2008 by several inventors. The differences are considerable and widely varied; however, the basis for deployment relies on one or more of the basic kinematic expressions previously described.

US Patent No.	Year	Inventor	Assignee	Title	Description
5394326	1995	Liu	Delco Electronic Corporation	Air bag deployment control system and method	ΔV + acceleration
5430649	1995	Cashler	Delco Electronic Corporation	SIR deployment method based on occupant displacement and crash severity	ΔV + jerk + displacement + acceleration
5587906	1996	McIver	TRW Inc.	Method and apparatus for sensing a vehicle crash condition using velocity enhanced acceleration crash metrics	ΔV + acceleration
5668720	1997	Takahashi	Toyoda Gosei Co., Ltd.	Air bag controlling apparatus	ΔV + jerk + acceleration
5777225	1998	Sada	Sensor Technology Co.	Crash sensor	ΔV + jerk + displacement + acceleration
5835007	1998	Kosiak	Delco Electronic Corporation	Method and apparatus for crash sensing using anticipatory sensor inputs	ΔV + acceleration
5948032	1999	Huang	Ford Global Technologies	Polynomial windowing algorithm for impact responsive activation	ΔV + jerk + displacement + energy
5999871	1999	Liu	Delphi Technologies	Control method for variable level airbag inflation	ΔV + jerk
6236921	2001	McConnell	Visteon Global Technologies	Three Speed Algorithm for Airbag Sensor Activation	ΔV + jerk + displacement
7424354	2008	Shen	Delphi Technologies	Supplemental restraint deployment method using dynamic crash classification	ΔV + jerk + displacement

Table 2.2 Algorithm Patents and Criteria

"Smart" Air Bags

Since air bags became a requirement for automakers in the late 1990's, there were occasions that air bags caused fatalities, especially to unrestrained, out-of-position children, in relatively low speed crashes. As a result, NHTSA updated *Federal Motor Vehicle Safety Standards (FMVSS) No. 208 - Occupant Crash Protection*, by adding a wide variety of new requirements, test procedures, and injury criteria, using an assortment of new dummies, and replacing the sled test with a rigid barrier crash test. Advanced air bag technology is also part of the design requirement. In response, U.S. automakers have introduced a new generation of advanced air bags, which implement multi-stage inflators to minimize the risk of injuries. Phase-in of the rule began with model year 2004 passenger vehicles. Full implementation of Phase 1 of this rule was required for all passenger vehicles by model year 2007.

These advanced occupant restraint systems are sometimes referred to as "smart" air bags because these systems can adapt their deployment strategies to the occupant belt status, occupant seating position, crash severity, and other factors. The system may choose to trigger the frontal air bags at different times and with a differing number of stages in order to optimize air bag performance (Gabler, et al. 2008).

The feature of dual-stage inflators allows two levels of inflation. Typically, in lower severity crashes, only the first-stage inflator would fire to reduce the risk of air bag induced injuries. In collisions in which the occupants are not belted, and also some severe crashes, both inflators would deploy to ensure that the air bags are fully inflated by the time the forward occupants reach their outer surface. Dual-stage air bag systems have improved performance over the baseline single-stage systems in terms of providing high-speed protection while reducing aggressivity to out-of-position occupants (Hollowell, et al. 2001).

References

- A. Alrabady & S. Mahmud. (2008). *Development of a Decision Making Algorithm for Airbag Control.* Wayne State University Press.
- D. Breed, W. Johnson & W. DuVall. (2000). *Vehicle Occupant Sensing System Including a Distance-Measuring Sensor on an Airbag Module or Steering Wheel Assembly.* U.S. Patent 6254127 B1.
- H. Gabler, C. Hampton & J. Hinch. (2004). *Crash Severity: A Comparison of Event Data Recorder Measurements with Accident Reconstruction Estimates.* SAE #2004-01-1194.
- H. Gabler & J. Hinch. (2008). *Evaluation of Advanced Air Bag Deployment Algorithm Performance Using Event Data Recorders.* Annals of Advances in Automotive Medicine 52.
- W. Hollowell, L. Summers, A. Prasad, G. Narwani & T. Ato. (2001). *Performance Evaluation of Dual Stage Passenger Air Bag Systems.* SAE #2001-06-0190.
- M. Huang. (2002). *Vehicle Crash Mechanics.* CRC Press.
- A. Hussain, M. Hannan, A. Mohamed, H. Sanusi & B. Majlis. (2006). *Decision Algorithm for Smart Airbag Deployment Safety Issues.* International Journal of electrical and Computer Engineering.
- K. Klaber. (1981). *Advanced Automotive Crash Recorder Design Development and Test Analysis.* SAE# 810809.
- W. Lee & I. Han. (2004). *Development of an Event Data Recorder and Reconstruction Analysis.* SAE #2004-01-1180.
- D. McConnell. (2001). *Three Speed Algorithm for Airbag Sensor Activation.* U.S. Patent 6236921 B1.
- National Highway Traffic Safety Administration. (2001, November). *Effectiveness of Occupant Protection Systems and Their Use – DOT HS 809 442.*

- J. Rowell. (2012). *Electronic Crash Sensing In Air Bag Litigation.* Retrieved from: http://www.hg.org/article.asp?id-18964.
- N. Takubo, T. Hiromitsu, K. Kato, K. Hagita, R. Oga & M. Kihara. (2011). *Study on Characteristics of Event Data Recorders in Japan; Analysis of J-NCAP and Thirteen Crash Tests.* SAE #2011-01-0810.

CHAPTER 3

DEFINING THE THRESHOLDS

According to the NHTSA (NHTSA 2003, p.5), "Air bags are typically designed to deploy in frontal and near-frontal collisions, which are comparable to hitting a solid barrier at approximately 8 to 14 mph." Specific thresholds are calibrated by each manufacturer according to vehicle size and stiffness. When the pre-defined threshold condition is met in frontal collisions, the ACM will issue the system "wake-up" or "algorithm-enable" command and begin the deployment decision-making algorithm. For frontal events, the pre-defined "wake-up" or "algorithm-enable" threshold is vehicle and module specific. For smaller vehicles, or vehicles with a lesser stiffness value, the deceleration threshold can be as low as 1.0 g; for larger vehicles, or vehicles with a greater stiffness value, the deceleration threshold may go up to 2.0 g's (Collision Safety Institute, 2011, p.44).

Essentially, when the ACM monitors the acceleration reading, it will make sure that the deceleration "jerk" (i.e., the change between two consecutive Δv samples) remains within the pre-defined threshold value; in the event that the sampled data presents a deceleration "jerk" beyond the threshold value, the ACM will proceed to the "wake-up" or "algorithm-enable" command, and a decision will be made to either fire the air bag or return to normal state.

Due to the proprietary nature of air bag deployment algorithms, the pre-defined threshold conditions for air bag deployment during a collision are not easily obtained; however, using the NHTSA guideline for an air bag to deploy in frontal barrier collisions within impact speeds of 8 to 14 mph, a range of threshold values can be estimated using known vehicle stiffness-to-weight ratios.

Threshold Estimates

In a collision, the amount of crush (C, in inches) at a given impact speed (V, in mph) is related to the ratio of the stiffness of a vehicle (k, in lb./in) and the vehicle weight (w, in lb.) by the following equation (Huang, 2002, Section 4.5.1):

$$\frac{C}{V} = 0.9\sqrt{\frac{w}{k}}$$

The time from the beginning of the impact to the time of the maximum crash pulse (t_m), in milliseconds, is:

$$t_m = 56.8\frac{C}{0.64V}$$

By substituting for C/V, the time (t_m) can be calculated using the weight-to-stiffness ratio as follows:

$$t_m \approx 80\sqrt{\frac{w}{k}}$$

Vehicle stiffness (k) can be determined from collision test results, which report mass (m), crush (c), and impact velocity (v) for vehicles subjected to frontal rigid barrier collision testing. Vehicle stiffness is calculated as:

$$k = \frac{mv^2}{2c^2}$$

The NHTSA website provides a large database of vehicle crash tests. For example, Test No. 6755 documents a frontal impact test of a 2010 Ford Fusion at an impact speed of 35.0 mph. The actual crush profile is measured at six crush zones. With these provided information, the vehicle stiffness (k) value for a 2010 Ford Fusion is calculated at approximately 3839 lb./in. The time to achieve the maximum crash pulse (t_m) is calculated at approximately 78 milliseconds. If we set the delta-v value at the NHTSA-suggested value for the barrier speed (8 to 14 mph), the corresponding deceleration value would be 4.7 to 8.2 g's.

Table 3.1 shows the corresponding range of decelerations and displacements in frontal barrier collisions, at which air bags are designed to deploy, given the calculated time to maximum crash pulse and different vehicle stiffness-to-weight ratios.

There is no significant correlation between vehicle weight and stiffness (G. Nusholtz, 1999). Two vehicles of similar weight may have very different stiffness values, as seen when comparing the 2010 Ford Fusion to a 2010 Toyota Prius. Both vehicles have approximately the same vehicle weight, yet the front-end stiffness of the Toyota Prius (5672 lb./in) is substantially greater than the Ford Fusion (3839 lb./in). Since both the amount of displacement and the duration of impact for a Ford Fusion are greater, an air bag would need to deploy in the Ford Fusion within a range of deceleration values lower than those required for the Toyota Prius.

Vehicle Year	Vehicle Model	Vehicle Class	Vehicle Weight w (lbs)	Vehicle Stiffness k (lbs/in)	Ratio k/w (lb/in/lb)	Time to max. pulse (ms)	Deceleration Range (g's)		Displacement Range (in)	
							(g's)	(g's)	(in)	(in)
2010	Ford Fusion	midsize	3640	3839	1.055	78	4.7	8.2	7.0	12.3
2010	Lexus RX350	SUV	4747	5690	1.199	73	5.0	8.7	6.6	11.5
2009	Ford Escape	SUV	4198	5070	1.208	73	5.0	8.8	6.6	11.5
2010	Honda Insight	compact	3115	3917	1.257	71	5.1	9.0	6.4	11.2
2010	Toyota Prius	midsize	3499	5672	1.621	63	5.8	10.2	5.7	9.9
2003	Infiniti G35	luxury	3352	11219	3.347	44	8.3	14.6	3.9	6.9
2007	Cadillac Escalade	SUV	5665	8970	1.583	63	5.7	10.0	5.7	10.0
2007	Ford Edge	SUV	4082	7993	1.958	57	6.4	11.2	5.1	9.0
1999	GMC Jimmy	SUV	3510	4261	1.214	72	5.0	8.8	6.5	11.4
2008	Scion tC	compact	2937	5401	1.839	59	6.2	10.8	5.3	9.3
2012	Ford F550	pickup	6832	8686	1.271	71	5.1	9.0	6.4	11.2
2008	Cadillac CTS	luxury	4101	8817	2.150	54	6.7	11.7	4.9	8.6
2007	Chevrolet Equinox	SUV	3172	8647	2.726	48	7.5	13.2	4.4	7.6
2007	Chevrolet Corvette	sportscar	3132	5526	1.764	60	6.1	10.6	5.4	9.5

Table 3.1 Air bag deployment ranges based on a barrier impact speed of 8 to 14 mph

References

- Collision Safety Institute. (2011). *Bosch Crash Data Retrieval System – Crash Data Retrieval Data Analyst Course Manual*.
- M. Huang. (2002). *Vehicle Crash Mechanics*. CRC Press.
- National Highway Traffic Safety Administration. (2003). *What You Need to Know About Air Bags – DOT HS 809 575*.
- National Highway Traffic Safety Administration. Databases and Software. Available at http://www.nhtsa.gov/Research/Databases+and+Software.
- G. Nusholtz, L. Xu, Y. Shi, L. DeDomenico. (1999). *Vehicle Mass, Stiffness and Their Relationship*. Daimler Chrysler Corporation Paper 05-0413.

CHAPTER 4

COMPARING VALUES

Real-world crashes are often not identical to solid barrier crashes, and care should be taken when comparing ranges of tested and calculated values. Impact duration does not vary significantly with impact velocity but varies greatly with the type of collision. For lower speed collisions, research (Anderson et al., 2015; Croft et al., 2000; Cipriani, 2002) indicates that impact duration generally ranges between 0.134 and 0.195 seconds (Tables 4.1, 4.2, and 4.3 below). When we examined the crash data (retrieved from Event Data Recorders) of vehicles involved in collisions at 35 mph or more, the impact duration falls near 0.120 seconds.

What follows in this chapter is a number of actual case studies to compare the actual recorded data with our estimated threshold. For each case, the data recorded by the EDR is downloaded through the Bosch Crash Data Retrieval (CDR) tool and then analyzed. Analysis of Crash Data Retrieval (CDR) data requires an exposure to and understanding of many potential system conditions and/or anomalies that may impact crash data (Collision 2011).

In the process of determining whether or not the air bag should deploy, we do look at the maximum Δv (delta-v, change of velocity) recorded by the air bag control module; yet, whether or not the maximum Δv falls within the barrier impact speed range of 8 to 14 mph is not the primary factor in our approach. First, we pinpointed the exact moment at which the deployment command, if applicable, was issued; next, we examined the data immediately before such a command, calculated the corresponding deceleration, and then compared the result with our estimated threshold value from the previously-discussed mathematical modeling to see whether the given condition warrants an air bag deployment and whether the air bag did, indeed, deploy.

Test	Impact Speed (mph)	"Engagement Time" (sec)
A1	2.36	0.211
A2	2.55	0.205
A3	2.36	0.212
A4	2.55	0.214
A5	2.55	0.219
A6	2.55	0.222
B1	5.10	0.204
B2	5.10	0.193
B3	5.03	0.205
B4	5.03	0.203
B5	5.10	0.194
B6	5.10	0.195
C1	7.95	0.170
C2	8.02	0.169
C3	8.02	0.159
C4	7.95	0.161
C5	7.95	0.202
C6	7.89	0.175
	Average Time =	0.195

Table 4.1 "Engagement Times" from Research by Anderson et al.

Test	Impact Speed (mph)	"Engagement Time" (sec)
S99-1	5.4	0.132
S99-2	9.3	0.154
S99-3	9.9	0.154
S99-6	6.6	0.182
S99-7	4.1	0.240
S99-9	7.0	0.220
	Average Time =	0.180

Table 4.2 "Engagement Times" from Croft's Machine vs. Man

Test	Impact Speed (mph)	"Engagement Time" (sec)
1	2.01	0.104
2	1.79	0.137
3	3.58	0.124
4	6.49	0.101
5	9.17	0.130
6	10.29	0.101
7	11.18	0.098
8	2.01	0.119
9	4.70	0.113
10	5.59	0.108
11	2.46	0.155
12	3.80	0.155
13	6.93	0.150
14	8.95	0.119
15	10.96	0.121
16	1.79	0.167
17	3.80	0.176
18	6.49	0.190
19	8.50	0.128
20	9.40	0.115
21	12.97	0.102
22	1.79	0.171
23	3.80	0.215
24	6.04	0.166
25	8.28	0.104
26	10.29	0.107
27	11.63	0.100
28	8.28	0.128
29	10.29	0.167
30	12.97	0.135
	Average =	0.134

Table 4.3 "Engagement Times" from Research by Cipriani et al.

Case Study #1 – 2007 Cadillac Escalade

Case Study #1 is a textbook example of air bag deployment triggered by a significant frontal impact. A 2007 Cadillac Escalade is traveling downhill; it moves into the opposite lane and collides head-on with an oncoming 2001 Nissan Frontier. The air bags of the 2007 Cadillac Escalade have deployed.

The crash data retrieved from the 2007 Cadillac Escalade indicates the initial contact at $t = -20$ msec. The maximum Δv of 35.01 mph is reported at $t = 100$ msec. The first-stage deployment command is issued at $t = 5$ msec. The cumulative Δv, prior to that, is last recorded as 4.46 mph at $t = 0$ msec. The deceleration immediately prior to the command ($t = -10$ to 0 msec) is approximately 11.6 g's. (It is noted that the driver's belt switch circuit shows a status of "unbuckled." A second-stage deployment command is issued at $t = 12.5$ msec. The corresponding deceleration immediately prior to the second-stage command is 17.4 g's.)

Utilizing our approach developed in Chapter 3, for the 2007 Cadillac Escalade, the expected barrier impact speed range of 8.0 to 14.0 mph for air bag deployment is equivalent to a deceleration range of 5.7 to 10.0 g's.

In this case, the driver's air bag of the 2007 Cadillac Escalade decides to deploy as soon as it detects the sudden deceleration (11.6 g's) beyond our estimated range (5.7 to 10.0 g's), despite the fact that the last recorded cumulative Δv is only 4.46 mph. Given the driver's unbelted status and the increasing deceleration, the second-stage deployment is warranted.

Air bag(s) may deploy as soon as the necessary criteria (i.e., a sudden substantial decrease in the change of velocity) is met.

Figure 4.1   2007 Cadillac Escalade (Air bag deployment)

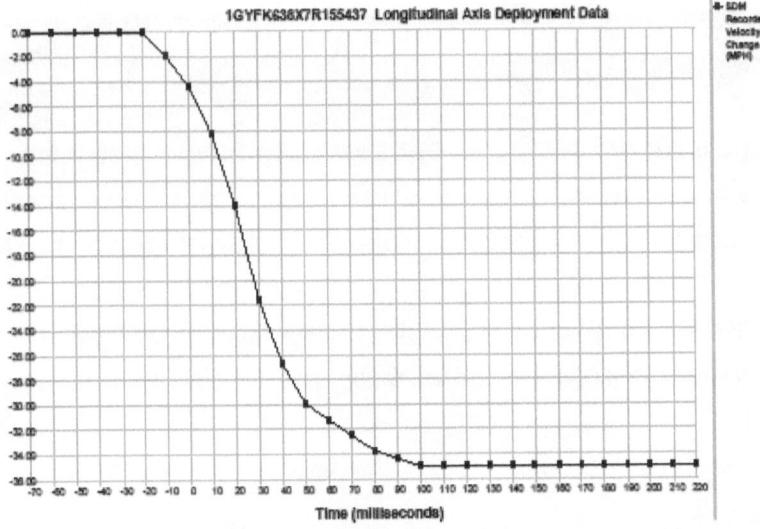

Figure 4.2   Recorded Δv data of the 2007 Cadillac Escalade

Case Study #2 – 2007 Ford Edge

Case Study #2, another textbook example of air bag deployment, involves a 2007 Ford Edge that experienced a significant frontal collision.

The crash data retrieved from the 2007 Ford Edge indicates the initial contact at $t = -15$ msec. The maximum Δv of 26.63 mph is reported at $t = 100$ msec. The first-stage deployment command is issued at $t = 13.5$ msec. The cumulative Δv, prior to that, is last recorded as 5.30 mph at $t = 13$ msec. In the 10 msec prior to $t = 13$ msec, the deceleration is 10.0 g's. (It is noted that the driver's belt switch circuit shows a status of "buckled." A second-stage deployment command is issued at $t = 23.5$ msec. The corresponding acceleration immediately prior to the second-stage command is 19.5 g's.)

Utilizing our approach developed in Chapter 3, for the 2007 Ford Edge, the expected barrier impact speed range of 8.0 to 14.0 mph for air bag deployment is equivalent to a deceleration range of 6.4 to 11.2 g's.

In this case, the driver's air bag of the Ford Edge decides to deploy as soon as the initial triggering deceleration (10.0 g's) falls within the range for air bag deployment (6.4 to 11.2 g's), despite the fact that the recorded cumulative Δv is only 5.30 mph at the time.

Among the kinematic variables, the deceleration "jerk" is a better and more reliable factor in the criteria for air bag deployment than the change of velocity.

Figure 4.3   2007 Ford Edge (Air bag deployment)

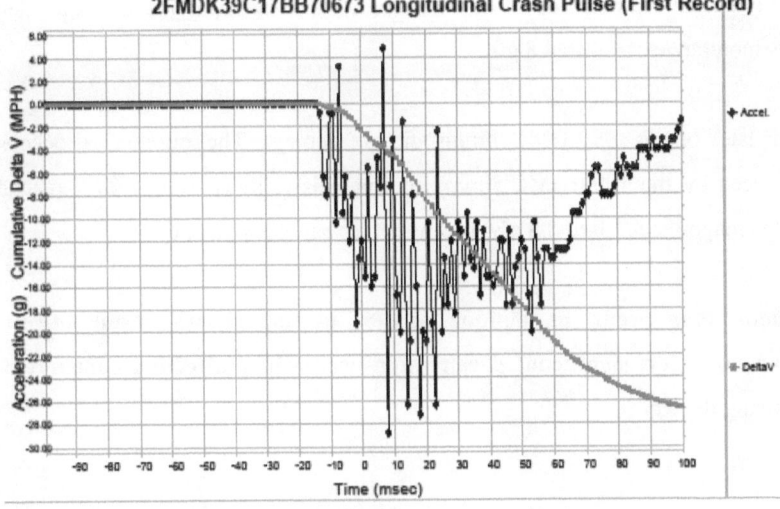

Figure 4.4   Recorded Δv data of the 2007 Ford Edge

Case Study #3 – 1999 GMC Jimmy

Air bag deployment is not always granted in a serious accident. In Case Study #3, a 1999 GMC Jimmy traveling on a sandy path, lost control, spun out, overturned, and rolled over. The vehicle underwent a series of violent movements and sustained quite a bit of deformation damage over its entire body; yet, there was no air bag deployment.

The air bag control module in the 1999 GMC Jimmy, in its course of monitoring the vehicle's activity, did not pick up any significant Δv in the longitudinal direction. The lack of frontal crush damage would be consistent with the fact that there was no significant longitudinal deceleration. It is noted that the driver's belt switch circuit shows a status of "buckled."

Utilizing our approach developed in Chapter 3, for the 1999 GMC Jimmy, the expected barrier impact speed range of 8.0 to 14.0 mph for air bag deployment is equivalent to a deceleration range of 5.0 to 8.8 g's.

The air bags of the 1999 GMC Jimmy did not deploy. The longitudinal deceleration experienced by the 1999 GMC Jimmy, during this rollover event, never reached the expected magnitude (5.0 to 8.8 g's) to warrant air bag deployment.

Collisions that occur at oblique angles do not always result in air bag deployment unless significant deceleration occurs in a direction concurrent with the sensing device.

Figure 4.5   1999 GMC Jimmy (No air bag deployment)

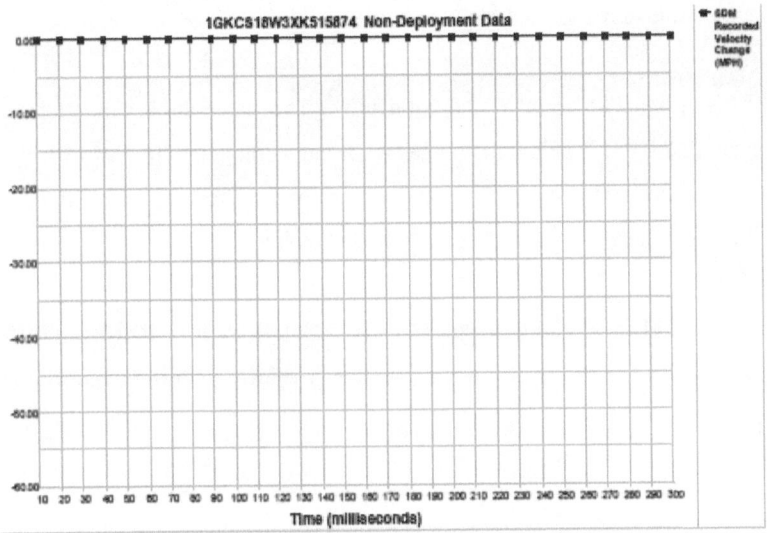

Figure 4.6   Recorded Δv data of the 1999 GMC Jimmy

Case Study #4 – 2008 Scion tC

Case Study #4 involves a vehicle-pedestrian collision; a 2008 Scion tC struck a pedestrian while it was traveling at approximately 50 mph on a freeway. Upon impact, the pedestrian went over the hood, caused some roof crush, and went through the front windshield.

The crash data retrieved from the 2008 Scion tC indicates the initial contact at $t = 0$ msec. The maximum Δv of 2.8 mph is reported at $t = 150$ msec. No deployment command is issued. Screening through the recorded data, the maximum deceleration is 2.73 g's. It is noted that the driver's belt switch circuit shows a status of "buckled."

Utilizing our approach developed in Chapter 3, for the 2008 Scion tC, the expected barrier impact speed range of 8.0 to 14.0 mph for air bag deployment is equivalent to a deceleration range of 6.2 to 10.8 g's.

The air bags of the 2008 Scion tC did not deploy. The maximum deceleration recorded in this event (2.73 g's) is below the range for air bag deployment (6.2 to 10.8 g's). It is our belief that the low deceleration experienced by the 2008 Scion tC is due to the weight difference between the pedestrian and the Scion tC.

Air bags do not always deploy in collisions in which there is a significant difference in weight between the bullet and target, such as a vehicle-pedestrian collision.

Figure 4.7   2008 Scion tC (No air bag deployment)

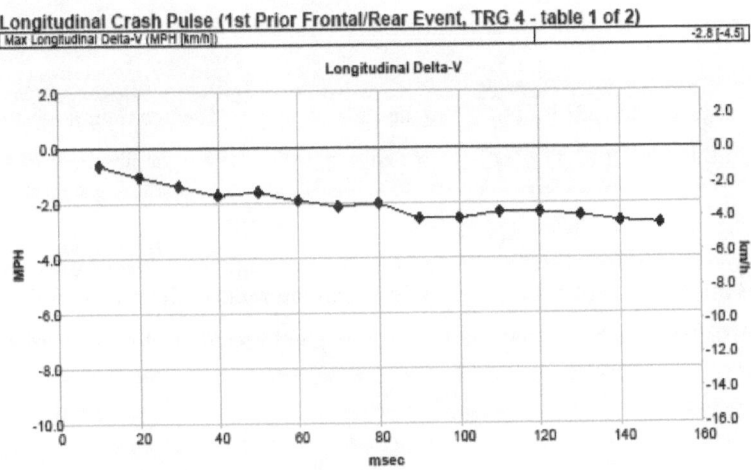

Figure 4.8   Recorded Δv data of the 2008 Scion tC

Case Study #5 – 2012 Ford F550

Case Study #5 involves a 2012 Ford F550 that appears to have sustained quite a bit of frontal collision damage. The 2012 Ford F550 is involved in a broadside collision in which the front end of the 2012 Ford F550 collides into the driver side of a 2003 Lexus RX300.

The crash data retrieved from the 2012 Ford F550 indicates the initial contact at $t = 0$ msec. The maximum Δv of 3.28 mph is reported at $t = 290$ msec. No deployment command is issued. Screening through the recorded data, the maximum deceleration is 2.37 g's. It is noted that the driver's belt switch circuit shows a status of "buckled."

Utilizing our approach developed in Chapter 3, for the 2012 Ford F550, the expected barrier impact speed range of 8.0 to 14.0 mph for air bag deployment is equivalent to a deceleration range of 5.1 to 9.0 g's.

The air bags of the 2012 Ford F550 did not deploy. The maximum deceleration recorded in this event (2.37 g's) is below the range for air bag deployment (5.1 to 9.0 g's).

Air bags might not deploy in collisions in which the relative degrees of stiffness are vastly different, such as the front of a vehicle impacting the side of another vehicle.

Figure 4.9   2012 Ford F550 (No air bag deployment)

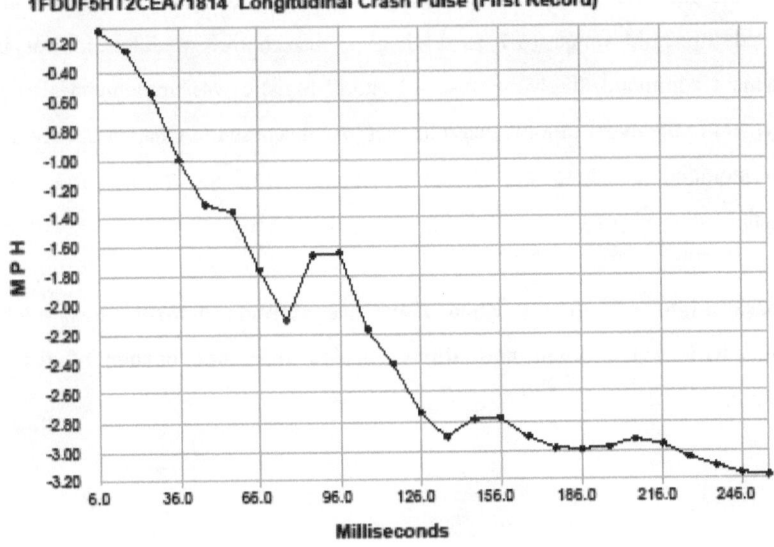

Figure 4.10 Recorded Δv data of the 2012 Ford F550

Case Study #6 – 2008 Cadillac CTS

Case Study #6 involved a 2008 Cadillac CTS that sustained a substantial amount of crush damage to its right front corner; however, the air bags of the 2008 Cadillac CTS did not deploy.

The crash data retrieved from the 2008 Cadillac CTS reported a maximum Δv of 21.80 mph at $t = 270$ msec. Screening through the recorded data, the maximum deceleration was as high as 11.6 g's. It is noted that the driver's belt switch circuit shows a status of "buckled."

For the 2008 Cadillac CTS, the expected barrier impact speed range of 8.0 to 14.0 mph for air bag deployment would be equivalent to a deceleration range of 6.7 to 11.7 g's.

Despite the fact that the recorded maximum deceleration (11.6 g's) fell within the estimated threshold range (6.7 to 11.7 g's), the control module did not issue a deployment command, likely because a "smart" logistic was implemented to prevent deployment in the event of a localized impact. As discussed in Chapter 2, recent air bag control modules may factor in *occupant sensing* and analysis of the *distance* from the occupant.

Air bags might not deploy when there are extreme deformations, such as a collision with a telephone pole during which only one corner of the car is deformed.

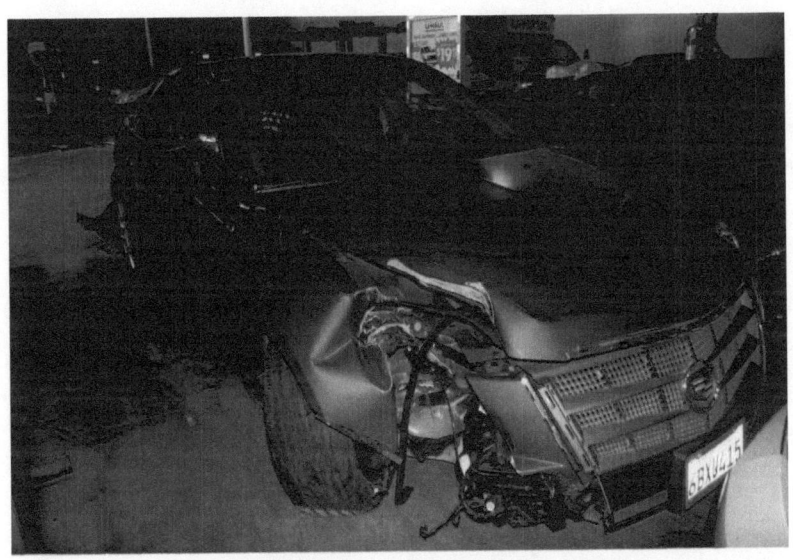

Figure 4.11 2008 Cadillac CTS (No air bag deployment)

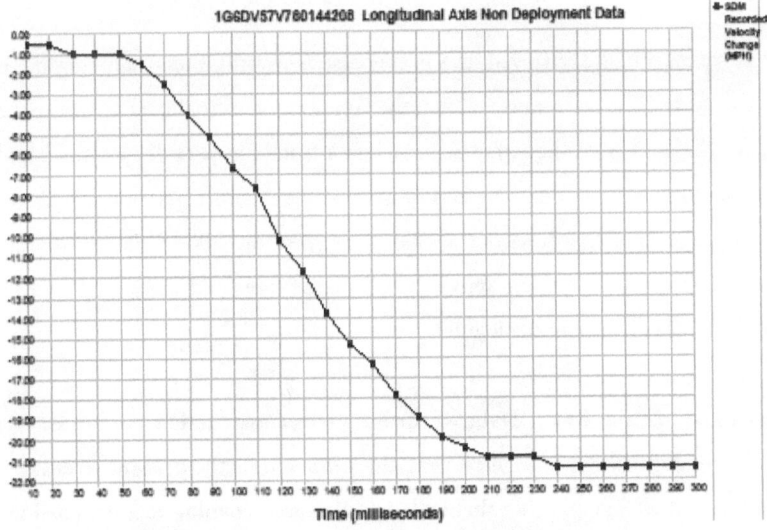

Figure 4.12 Recorded Δv data of the 2012 Ford F550

Case Study #7 – 2007 Chevrolet Equinox

We revisited a case study (Case Study #7) that was cited in our previously published article – a 2007 Chevrolet Equinox collided with a Harley-Davidson motorcycle. The air bags of the 2007 Chevrolet Equinox are not deployed in the collision.

The crash data retrieved the 2007 Chevrolet Equinox, for unknown reason, starts at $t = -70$ msec with a maximum Δv of 9.27 mph from $t = -70$ to 110 msec. The reading then returns to a steady 8.55 mph from $t = 120$ msec and on. It is noted that the driver's belt switch circuit shows a status of "buckled."

For the 2007 Chevrolet Equinox, the expected barrier impact speed range of 8.0 to 14.0 mph for air bag deployment would be equivalent to a deceleration range of 7.5 to 13.2 g's.

For unknown reason, the retrieved data in this case is invalid, because it does not exhibit common features of a crash pulse. First, the Δv monitored by the EDR is a cumulative value and must start from the value of 0. When a frontal impact occurs, the cumulative Δv increases in one direction. The end of cumulative changes generally marks the end of the crash event. Regardless, we would not expect the 2007 Chevrolet Equinox to have experienced a deceleration above the estimated threshold range (7.5 to 13.2 g's), because there was no air bag deployment.

In order to utilize the Event Data Recorder for reconstruction, one must examine the recorded data and make sure that the data correctly present a crash event, rather than just cherry-picking the peak Δv value and rushing to a conclusion.

Figure 4.13 2007 Chevrolet Equinox (No air bag deployment)

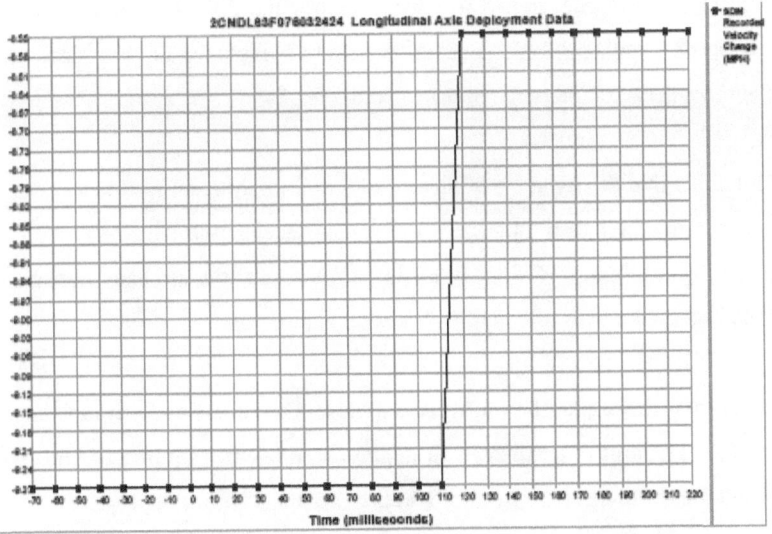

Figure 4.14 Recorded Δv data of the 2007 Chevrolet Equinox

Case Study #8 – 2007 Chevrolet Corvette

Case Study #8 is another case study that was cited in our previously published article – a 2007 Chevrolet Corvette struck several small signs, trees, and a utility pole off road at a very high rate of speed. The air bags deployed.

The crash data retrieved from the 2007 Chevrolet Corvette indicates the initial contact at $t = 10$ msec. The maximum Δv of 9.33 mph is reported at $t = 110$ msec. The deployment command, however, is issued at $t = 32.5$ msec. As we examined the recorded data immediately prior to the deployment command being issued, we saw a change in the Δv of 2.48 mph. The corresponding deceleration is calculated at 11.3 g's. It is noted that the driver's belt switch circuit shows a status of "unbuckled."

Utilizing our approach developed in Chapter 3, for the 2007 Chevrolet Corvette, the expected barrier impact speed range of 8.0 to 14.0 mph for air bag deployment is equivalent to a deceleration range of 6.1 to 10.6 g's.

Despite the fact that the Δv at $t = 30$ msec (2.79 mph) may seem well below the typical deployment range of 8.0 to 14.0 mph, the actual deceleration (11.3 g's) is above the estimated threshold range (6.1 to 10.6 g's); as a result, the air bags deployed early enough and saved the lives of both the passenger and driver.

In some events, there may be multiple impacts. One must understand the correct moment that air bags are expected to deploy, so one can examine the full scope of the recorded data in order to predict whether air bags would deploy.

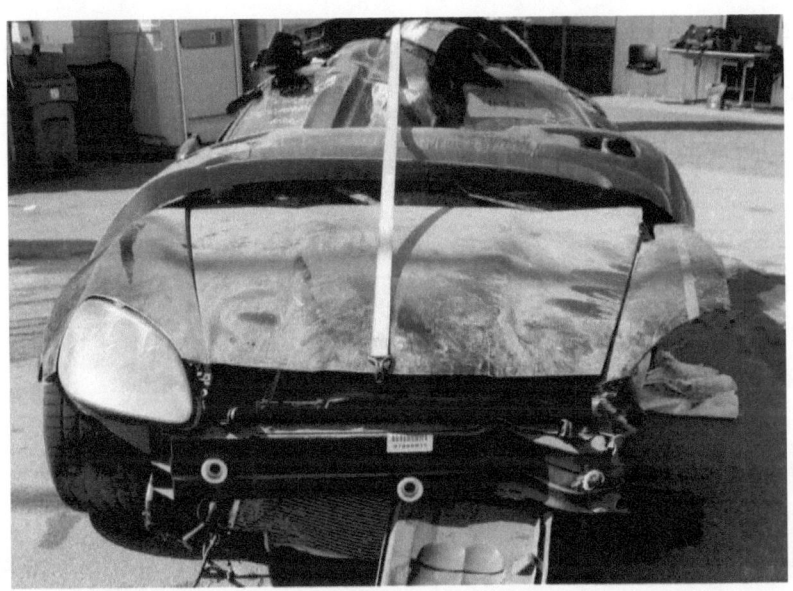

Figure 4.15 2007 Chevrolet Corvette (Air bag deployment)

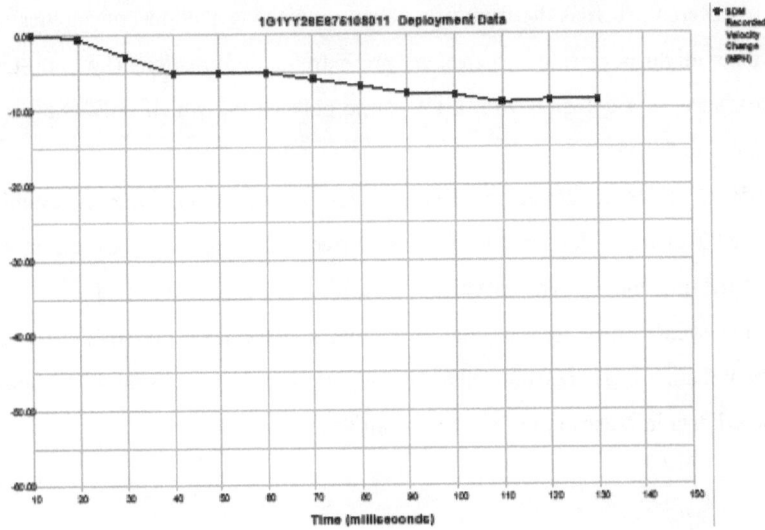

Figure 4.16 Recorded Δv data of the 2007 Chevrolet Corvette

Summary

After comparing our prediction with actual case studies, we noted the following:

- Air bag(s) may deploy as soon as the necessary criteria (i.e., a sudden substantial decrease in the change of velocity) is met;
- Among the kinematic variables, the deceleration "jerk" is a better and more reliable factor in the criteria for air bag deployment than the change of velocity;
- Collisions that occur at oblique angles do not always result in air bag deployment unless significant deceleration occurs in a direction concurrent with the sensing device;
- Air bags do not always deploy in collisions in which there is a significant difference in weight between the bullet and target, such as a vehicle-pedestrian collision;
- Air bags might not deploy in collisions in which the relative degrees of stiffness are vastly different, such as the front of a vehicle impacting the side of another vehicle;
- Air bags might not deploy when there are extreme deformations (i.e., non-uniform crush), such as a collision with a telephone pole during which only one corner of the car is deformed;
- In order to utilize the Event Data Recorder for reconstruction, one must examine the recorded data and make sure that the data correctly present a crash event, rather than just cherry-picking the peak Δv value and rushing to a conclusion; and
- In some events, there may be multiple impacts. One must understand the correct moment that air bags are expected to deploy, so one can examine the full scope of the recorded data in order to predict whether air bags would deploy.

References

- R. Anderson, J. Welcher, T. Szabo, J. Eubanks & W. Haight. (1998). *Effect of Braking on Human Occupant and Vehicle Kinematics in Low Speed Rear-End Collisions.* SAE #980298.
- A. Cipriani, F. Bayan, M. Woodhouse, A. Cornetto, A. Dalton, C. Tanner, T. Timbario & E. Deyeri. *Low Speed Collinear Impact Severity: A Comparison between Full Scale Testing and Analytical Prediction Tools with Restitution Analysis.* SAE #2002-01-0540.
- Collision Safety Institute. (2011). *Bosch Crash Data Retrieval System.* Data Analyst Course Manual.
- A. Croft. (2000). *Machine vs. Man – Human Subjects Crash Testing.* DVD Video from CRASH 1999-2000.
- M. Varat & S. Husher. (2010). *Crash Pulse Modeling for Vehicle Safety Research.* KEVA Engineering Paper 501.

CHAPTER 5

CONCLUSION

Air bags are designed to be a Supplemental Restraint System (SRS), mostly to enhance protection for belted occupants in frontal crashes. Air bags must be able to recognize a crash correctly and inflate early enough, in order to provide the desired cushion for the occupants and mitigate secondary impact(s) between the occupants and the interior of the vehicle.

The air bag control module (ACM) continuously monitors the acceleration data reported by the crash sensors; when the pre-defined threshold condition is met, the ACM issues the system "wake-up" or "algorithm-enable" command and begins the deployment decision-making algorithm based on one or more kinematic variables (i.e., Δv and/or deceleration "jerk", etc.) as well as other inputs, such as occupant presence status, seat belt buckle status, etc.

Due to the proprietary nature of air bag deployment algorithms, the pre-defined threshold conditions for air bag deployment during a collision are not easily obtained. Given the NHTSA guideline that air bags are designed to deploy at a barrier impact speed of 8 to 14 mph, a range of deceleration and displacement threshold values can be calculated (based on vehicle stiffness-to-weight ratios) and used to study whether an air bag should or should not deploy in a collision. In this book, we have presented an approach to estimate the deceleration "jerk" threshold range, so a baseline is established when we review the actual data recorded.

Event Data Recorder (EDR), the "Black Box," may contain valuable information for automakers' research and reconstruction studies. In order to comprehend and analyze the crash data in the EDR, one must:

- Understand what air bags are designed to do;
- Understand the general steps involved in the deployment decision-making process;
- Be familiar with common factors constituting the air bag deployment criteria; and
- Have an exposure to and understanding of many potential system conditions and/or anomalies that may impact the crash data.

The presented case studies provide many interesting findings. In typical frontal crashes, air bags deploy in the early part of the event, rather than when the maximum Δv (change of velocity) is achieved. Among common kinematic variables, the deceleration "jerk" is more accurate and reliable as a factor in the criteria for air bag deployment than the Δv.

In three out of the eight presented case studies (Cases #1, #2, and #8), air bags have deployed; the corresponding deceleration "jerk" values are all within or beyond our estimated threshold ranges. Among the remaining five case studies (Cases #3, #4, #5,

#6, and #7), air bags are not deployed; three of them (including the rollover event in Case #3) can be explained as their deceleration "jerk" values being under our estimated threshold values.

The non-deployment of the 2008 Cadillac CTS in Case Study #6 is most likely due to a "smart" logistic implemented in its ACM, since the distance from the occupant indicates a localized, extreme corner collision and prevents deployment. In Case Study #7, the crash data retrieved from the EDR seem invalid, because it does not exhibit the common features of a crash pulse. Regardless, given the localized corner crush profile and the fact that the bullet vehicle (a motorcycle) has a much smaller weight, we can understand why the air bags are not deployed.

Overall, the presented approach of determining whether air bags should or should not deploy is primarily based on the deceleration "jerk" value, as well as other conditions (i.e., a distant and localized corner impact) unrelated to kinematic variables; the results from the case studies are reasonably satisfying. Therefore, with certain rules and exceptions in mind, this approach utilizing the deceleration "jerk" can be used as a useful forensic analysis tool in vehicle crashes.

APPENDIX

GROSSARY OF TERMS

(Partially extracted form Data Analyst Course Manual by Collision Safety Institute)

Acceleration: a vector quantity specifying the rate of change of velocity.

Accelerometer: a device which converts mechanical motion into an electrical signal proportional to the acceleration value of the motion; it measures inertial acceleration or gravitational force.

Air bag Control Module (ACM): the control module for air bags and related restraint systems.

Algorithm: a step-by-step set of operations designed to accomplish a specific task.

Algorithm enable (AE): a programmed threshold for a specific ACM at which the ACM begins the deployment decision making algorithm.

Control Module: an electronic device that makes decisions and controls other devices.

Crash Pulse: the period of time defined by the moment when two vehicles come into contact until the point where they separate at the centroid of damage and the exchange of momentum between the vehicles ends. As "crash duration," it is defined by time.

Delta-V: the vector change of speed of a vehicle involved in an "event" described by a magnitude and direction.

Delta-V, longitudinal: a cumulative change in velocity, as recorded by the EDR of the vehicle, along the longitudinal axis, starting from crash time zero and ending at 0.25 seconds, recorded every 0.01 seconds (NHTSA 49 CFR Part 563 definition).

Deployment (Event): acceleration observed along one of the car's axes sufficient to cause the control module's crash sensing algorithm to "enable" or "wake up" and is sufficient to warrant a commanded deployment.

Event Data Recorder (EDR): a function within an ACM, PCM or ROS (in the context of this version of CDR) that has the capability to save certain crash parameters after primary functions are completed. It may record a vehicle's dynamic, time-series data during and/or just prior to a crash event intended for retrieval after the crash event (may also be defined as part of NHTSA 49 CFR Part 563 in context).

Enabled: when a threshold has been met satisfying one of the criteria necessary to begin a process or deploy a device.

Event: the occurrence of some level of acceleration that causes an ACM to evaluate available data and decide whether or not to deploy restraint devices(s); a crash or other physical occurrence that causes a module's trigger threshold to be met or exceeded.

Frontal Air Bag: the primary inflatable occupant restraint device that is designed to deploy in a frontal crash to protect the front seat occupants. It requires no action by vehicle occupants and is used to meet the applicable frontal crash protection requirements of FMVSS No. 208.

G, g: unit of measurement designation for acceleration.

Jerk: the rate of change of acceleration.

Maximum delta-V, longitudinal: the maximum value of the cumulative change I velocity, as recorded by the EDR in the vehicle along the X-axis, starting from crash time zero and ending at 0.3 seconds (NHTSA 49 CFR Part 563 definition).

Millisecond: a millisecond is 0.001 seconds.

Non Deployment (event): acceleration observed along one of the car's axes sufficient to cause the module's crash sensing algorithm to "enable" or "wake up" but does not warrant a commanded deployment.

Safety Belt Status: also seen as seat belt buckle circuit status. This reflects the feedback from the safety system that is used to determine if an occupant's safety belt is buckled or not buckled (NHTSA 49 CFR Part 563 Definition).

SRS: Supplemental Restraint System

Trigger threshold: trigger threshold is the point at which a change in acceleration normally along the vehicle X axis equals or exceeds a pre-determined value as set in an ACM's calibration (NHTSA 49 CFR Part 563 Definition).

Wake-up: a programmed threshold for a specific ACM at which the ACM begins the deployment decision making algorithm. See also "algorithm enable."

ABOUT THE AUTHORS

Kenneth Alvin Solomon, Ph.D., P.E., Post Ph.D., retired from RAND as Senior Scientist after nearly 23 years of service. He served as faculty at The RAND Graduate School and Adjunct Faculty at UCLA, USC, The Naval Post Graduate School, and George Mason University, publishing approximately 200 peer-reviewed papers and fourteen prior books.

Wei-Kuang Chao, B.S., M.S., obtained a Bachelor of Science degree in Mechanical Engineering from Worcester Polytechnic Institute in Worcester, Massachusetts, and a Master of Science degree in Mechanical Engineering from Columbia University in New York, New York. With nearly 16 years of service at The Institute of Risk and Safety Analyses, Mr. Chao is currently a Senior Forensic Scientist specializing in accident reconstruction, biomechanics and human factors.

Mr. Jesse Kendall, P.E., obtained a Bachelor of Science degree in Civil Engineering from the University of Vermont in Burlington, Vermont. He completed his engineering internship in Denver, Colorado, where he worked for two prominent structural engineering consulting firms before becoming a licensed professional engineer in six states. Recently in California, he worked as a Forensic Scientist at the Institute of Risk and Safety Analyses. Mr. Kendall now enjoys dividing his time between his own engineering consulting business and producing Vermont maple syrup.

I want morebooks!

Buy your books fast and straightforward online - at one of the world's fastest growing online book stores! Environmentally sound due to Print-on-Demand technologies.

Buy your books online at
www.get-morebooks.com

Kaufen Sie Ihre Bücher schnell und unkompliziert online – auf einer der am schnellsten wachsenden Buchhandelsplattformen weltweit!
Dank Print-On-Demand umwelt- und ressourcenschonend produziert.

Bücher schneller online kaufen
www.morebooks.de

OmniScriptum Marketing DEU GmbH
Heinrich-Böcking-Str. 6-8
D - 66121 Saarbrücken
Telefax: +49 681 93 81 567-9

info@omniscriptum.com
www.omniscriptum.com

www.ingramcontent.com/pod-product-compliance
Lightning Source LLC
Chambersburg PA
CBHW031547210526
45464CB00003B/1186